FORESTRY COMMISSION BULLETIN 84

Taxation of Woodlands

LONDON: HER MAJESTY'S STATIONERY OFFICE

© *Crown copyright 1989*
First published 1989

ISBN 0 11 710274 1
ODC 95:(410)

Previous editions of this title have been issued as Forestry Commission Leaflet 12.

KEYWORDS: Forestry, Legislation, Woodlands

Taxation of Woodlands

This Bulletin has been written with the help of the Inland Revenue. It is for guidance only and reflects the tax position at the time of writing.

It should be borne in mind that these notes are not binding in law and in a particular case there may be special circumstances which will need to be taken into account.

They do not affect rights of appeal.

Enquiries relating to this publication
should be addressed to:
The Technical Publications Officer,
Forestry Commission, Forest Research Station,
Alice Holt Lodge, Wrecclesham,
Farnham, Surrey, GU10 4LH.

Contents

	Page
Introduction	v
Part 1 Income tax on woodlands	1
Part 2 Capital gains tax on the disposal of woodlands	4
Part 3 Inheritance tax (and capital transfer tax) on woodlands	6
Part 4 Estate duty on woodlands	11
Value added tax	11
Acknowledgements	12
Further reading	12

Impôts Forestiers

Sommaire

Ce Bulletin explique les impôts forestiers dans Le Royaume-Uni dans l'interêt des occupants des forêts (et leurs conseillers).

Part 1 traite des impôts sur le revenu et des impôts corporatifs, après les grands changements fiscaux par la Loi de Finance 1988 en ce qui concerne les forêts de production.

Part 2 traite des impôts en ce qui concerne les profits capitaux après les dispositions des forêts.

Part 3 traite des engagements en ce qui concerne des impôts de succession (et les impôts précédents sur le transfert du capital) pour les donations entre vifs après 26 Mars 1974, et pour les décès après 12 Mars 1975.

Part 4 traite des engagements en ce qui concerne les droits de succession pour les décès avant 13 Mars 1975.

Waldbesteuerung

Zusammenfassung

Dieses Bulletin erklärt die Waldbesteuerung im Vereinigten Königreich zum Nutzen von Waldinhabern (und ihren Beratern).

Abteil 1 behandelt Einkommensteuer und Körperschaftssteuer nach den gründlichen Änderungen in der Besteuerung von Erwerbswäldern, die durch das Finanzgesetz 1988 verursacht wurden.

Abteil 2 behandelt die Besteuerung der Kapitalgewinne bei der Übergabe von Wäldern.

Abteil 3 behandelt die Haftung für Erbschaftssteuer (und Kapitalnachlasssteuer, die dadurch ersetzt wurde) betreffs Schenkungen während der Lebenszeit des Schenkers, die nach 26 März 1974 gemacht wurden, und betreffs Tode nach 12 März 1975.

Abteil 4 behandelt die Haftung für Nachlasssteuer betreffs Tode vor 13 März 1975.

Taxation of Woodlands

Introduction

This Bulletin explains the taxation of woodlands for the benefit of woodland occupiers (and their advisers).

Part 1 deals with income tax and corporation tax following the major changes to the taxation of commercial woodlands made by the Finance Act 1988.

Part 2 deals with the application of capital gains tax to disposals of woodlands.

Part 3 deals with liability to inheritance tax (and capital transfer tax which it replaced) for lifetime gifts made after 26 March 1974 and for deaths after 12 March 1975.

Part 4 deals with liability to estate duty for deaths before 13 March 1975.

Any enquiry about income tax, corporation tax and capital gains tax should be addressed to the Inspector of Taxes to whom the woodland occupier normally makes his tax returns. Any enquiry about inheritance tax, capital transfer tax, or estate duty should be sent to the appropriate Capital Taxes Office (see 3.21, page 9).

Part 1
Income tax on woodlands

General

1.1 The Finance Act 1988 made significant changes in the tax treatment of commercial woodlands, that is, woodlands in the United Kingdom which are managed on a commercial basis and with a view to the realisation of profits. Previously, an occupier of commercial woodlands was liable to income tax under Schedule B on one-third of the annual value of the woodland (and there was no separate tax charge on profits from the sale of trees). Alternatively, the occupier could elect to be assessed on the profits from the sale of trees under Schedule D.

1.2 The Finance Act 1988 abolished the charge to tax under Schedule B (with effect from 6 April 1988) and removed the option to elect to be assessed under Schedule D (with effect from 15 March 1988). So income (and corporation) tax is no longer chargeable on the occupation of commercial woodlands, and expenditure on the cost of planting and maintaining commercial woodlands is not allowable as a tax deduction. Tax relief will no longer be available for any interest paid in connection with forestry activity, on a loan, for example, to buy woodlands or to invest in a company or partnership engaged in forestry activity. But there are special arrangements which will operate until 5 April 1993. These are described in paragraphs 1.4 onwards.

1.3 Rents and other revenue receipts (for example, periodic payments for a right of way) from woodlands, apart from profits from the sale of timber, etc., are assessable under the general rules for the taxation of income from land in the UK.

Transitional arrangements until 5 April 1993

1.4 There are transitional arrangements for those who were occupiers (typically owners or tenants) of commercial woodlands before 15 March 1988. They can elect to be assessed and charged to tax under Schedule D for the tax years up to 1992-93, and if they do so, they can also continue to offset the costs of planting and maintaining woodlands against taxable income for the same period.

1.5 The transitional arrangements are also available to:
- anyone who became an occupier of commercial woodlands after 15 March 1988 as a result of a firm commitment evidenced in writing and entered into before that date, or
- anyone who had made a grant application to the Forestry Commission for commercial woodlands before 15 March 1988.

1.6 The transitional arrangements are not available to an occupier who receives a grant under the new Woodland Grant Scheme. Grants paid under this scheme are free of income and corporation tax, and are set at levels which reflect the fact that occupiers can no longer claim tax relief for the costs of planting and maintenance.

The transitional provisions will not apply for any chargeable period following that in which a Woodland Grant Scheme grant is received. A chargeable period is the income tax year (6 April to the following 5 April) for individuals and the accounting year for companies. If, for any reason, the occupier has already had the benefit of tax relief for planting or other costs before the grant is paid, the Inspector will adjust the tax position once payment is actually made.

1.7 An occupier who had an election for Schedule D treatment in force at 15 March 1988, will automatically qualify for transitional relief for the woodlands covered by the election. Others who wish to benefit from the transitional arrangements must inform the Inspector of Taxes for the district in which the woodlands are situated in writing not later than 2 years after the end of the chargeable period to which the election relates.

The election must extend to **all** woodlands on the same estate managed on a commercial basis. But an occupier can treat woodlands which have been planted or replanted as being on a separate estate if he gives notice to the Inspector of Taxes within 2 years of the planting or replanting.

If the occupier does not wish this 'notice' to apply to all his woodlands planted within the last 2 years, he should give particulars.

In this way, if an occupier wishes to clear and replant storm damaged woodland in stages he can make a series of elections, one for each area as it is replanted under the Woodland Grant Scheme. This means that they will be treated as separate estates outside the income tax system and tax relief can continue for the clearance of the remainder of the woodland.

An election to be assessed under Schedule D is binding *for the remainder of the transitional period* as long as the woodlands are occupied by the person making the election.

Assessment under Schedule D during the transitional period

1.8 Once a Schedule D election has been made, profits from the occupation of commercial woodlands will be calculated as if they were profits of a trade and should include:

- all receipts from the sale of trees;
- grants from the Forestry Commission (other than those made under the Woodland Grant Scheme) — see paragraph 1.3;
- grants made by other government departments;
- any insurance money received as a result of damage to, or the destruction of, trees.

Insurance receipts of a capital nature should be left out of the Schedule D computation but an occupier may be liable to capital gains tax on them.

These rules also apply where commercial woodlands are occupied by a company chargeable to corporation tax.

Basis of assessment

The normal basis of assessment under Schedule D is the profit for the year immediately preceding the year of assessment. But in the case of newly planted or replanted woodlands treated as a separate estate (see paragraph 1.7) the computation of the profits should be made as if a new business had been set up, without taking into account any profits or losses arising from the land in question before the date of planting or replanting, as the case may be.

Date to which accounts may be made up

The income tax year begins on 6 April and ends on the following 5 April. A woodland occupier should therefore make up his accounts for tax purposes for a period of 12 months ending on a day within the tax year preceding the year of assessment. For example, the assessment for the year 1988-89 would be based on the profits of the year ended 5 April 1988, if the accounts are made up annually to 5 April. Or if the accounts are made up annually to 31 December, on the profits of the year ended 31 December 1987.

Form of accounts

The Inland Revenue will normally be prepared to accept accounts which are made up each year by reference to receipts and payments in the year (the 'cash' basis), instead of on the more accurate and correct basis of accrued receipts and expenses (the 'earnings' basis). The profits of the first 3 years from the date of commencement **must** be computed on the 'earnings' basis. A form of account for occupiers who want to change to a cash basis (Form 10W) can be obtained from the Inspector of Taxes.

Loss relief

An occcupier who has elected for the transitional arrangements and who makes a loss in any year from the occupation of his woodlands may claim to set that loss against his other income for the year of claim (or the following year). For this purpose, a loss may include an excess of capital allowances. Usually the capital allowances in question will be those due for the year following the year of loss.

Loss relief must be claimed by writing to the Inspector of Taxes within 2 years after the end of the year of assessment for which the relief is claimed.

Any company which has elected for Schedule D treatment under the transitional provisions may claim relief for any losses incurred in accordance with the corporation tax rules.

Relief for capital expenditure

During the transitional period an owner or tenant of forestry land who has elected to be assessed to income tax under Schedule D and who has incurred capital expenditure on:

- the construction or reconstruction of forestry buildings (including extensions and adaptations);
- cottages;
- fences;

or

- other work for the purposes of forestry on the land,

may be able to claim capital allowances.

Since 1 April 1986 this sort of capital expenditure, qualifies for annual writing down allowance of 4%. A system of balancing adjustments operates, at the taxpayer's request, when an asset is destroyed, demolished or sold. This enables the allowances to be brought into line with actual depreciation. Allowances for capital expenditure on plant and machinery used in the woodlands may be claimed on the same basis as for a trade.

Appeals

If an occupier of woodlands is dissatisfied with the amount of an assessment made under Schedule D, he may appeal in writing to Inspector of Taxes within 30 days from the date of issue of the notice of assessment. Usually appeals are settled by agreement with the Inspector. Otherwise the matter may be referred to the General or Special Commissioners for an independent ruling. Details of how the appeals procedure works can be found in leaflet IR37 *Appeals* available from any Tax Office.

Part 2
Capital gains tax on the disposal of woodlands

2.1 Capital gains tax is chargeable on any increase in value — after allowance for inflation — on the disposal of chargeable assets. Land is a chargeable asset for the purposes of the tax and 'disposal' includes any occasion on which the ownership of the asset is transferred, whether in whole or in part from one person to another (except on death), for example by sale, exchange or gift.

2.2 For individuals, chargeable gains are treated as the top slice of income and charged to capital gains tax at income tax rates. Unused income tax reliefs and allowances cannot be set against gains. Companies' gains are chargeable to corporation tax at normal corporation tax rates. Gains accruing to trustees are generally charged at the equivalent of the basic income tax rate although trustees of discretionary and accumulation trusts are liable at the equivalent of the sum of the basic plus additional rates. Special rules apply where the settlor or his/her spouse has an interest in the settlement. If in doubt, your Inspector of Taxes will be pleased to help.

2.3 A transfer of an asset by gift or otherwise than by sale at market value is generally deemed to take place for a consideration equal to the market value of the asset. The chargeable gains arising on such transfers may in some circumstances be deferred until a subsequent disposal of the asset.

Special rules for commercial woodlands

2.4 Special rules apply to the computation of gains and losses arising on the disposal of commercial woodlands in the United Kingdom. Their effect is to exclude growing timber (as distinct from the land on which it stands) from the computations. (Growing timber includes saleable underwood.) Thus the computation does not take any account of the price paid for the trees growing on the land at the time of the disposal or the cost of acquiring (or planting) and maintaining those trees. Capital expenditure on the construction, reconstruction, extension or adaptation of forestry buildings, cottages and fences and other capital improvements is deductible in arriving at the gain or loss on disposal. During the transitional period to 5 April 1993 any income tax allowances for such capital expenditure (see paragraph 1.8, page 3) are left out of account in the capital gains tax computation, unless the land is disposed of at a loss. If it is disposed of for less than the owner paid for it, the loss is not allowable for capital gains tax purposes to the extent that it has been covered by capital allowances.

Commercial woodland (but not growing timber) is a qualifying business asset for the purposes of capital gains tax rollover relief. This means that where all or part of the proceeds from disposal of other qualifying assets are reinvested in commercial woodland (or vice versa) within certain time limits, rollover relief may be available to defer all or part of the tax which would otherwise be payable immediately. Your Inspector of Taxes will explain how this relief works upon request.

Non-commercial woodlands

2.5 Woodlands which are not run on a commercial basis are subject to the normal capital gains tax rules. Felled trees are treated as chattels for capital gains tax purposes. A chargeable gain can arise only if an individual tree is sold for more than £3000.

Further information

2.6 Information on the general operation of capital gains tax is contained in the Board of Inland Revenue's leaflet CGT14 *Capital gains tax — an introduction,* copies of which may be obtained from any Tax Office. Queries about the application of the tax to a particular holding of woodlands should be made to the Inspector of Taxes for the District which deals with your income tax returns (or in the case of a company, the corporation tax return).

Part 3
Inheritance tax (and capital transfer tax) on woodlands

Inheritance tax and capital transfer tax — when chargeable

3.1 The Finance Act 1986 introduced inheritance tax for transfers made on or after 18 March 1986. Capital transfer tax, which in turn had replaced estate duty, applied to lifetime transfers made after 26 March 1974; and to transfers on death on or after 13 March 1975. Like its predecessor, inheritance tax applies to 'transfers of value' — broadly, transfers which reduce the value of the transferor's estate. Capital transfer tax was chargeable, subject to certain exemptions, on all lifetime transfers.

3.2 Under inheritance tax, most outright gifts become exempt from tax if the transferor survives 7 years from the date of the gift. During that 7-year period they are called 'potentially exempt transfers' (PETS). PETS may be made:

- by one individual to another individual; or
- by an individual into an accumulation and maintenance trust; or
- by an individual into a trust for the disabled.

A gift cannot be a PET so long as the transferor reserves or enjoys a benefit from the gift.

The amount charged and rates of tax

3.3 The rules for determining the amount charged for inheritance tax are the same as they were for capital transfer tax. The amount chargeable to tax on a lifetime gift is measured by the fall in the value of the estate as the result of the transfer. If the transferor, rather than the transferee, pays the tax, the amount of the tax has to be included in the fall in value. On death, tax is charged on the estate as valued immediately before that event. The estate includes settled property in which the deceased has an interest in possession.

Inheritance tax rates

3.4 The rate of tax on any transfer depends on the cumulative total of transfers (other than exempt transfers) made in the previous 7 years. This applies to transfers made on death as well as to lifetime transfers which are immediately chargeable (for example, transfers into discretionary trusts) or which become chargeable because the transferor dies within 7 years of making them. No tax is payable on the first part, currently £118 000, of the cumulative total. This threshold is increased from 6 April every year in line with the increase in the retail prices index for the year to the previous December, unless Parliament decides otherwise. Above the threshold, tax on transfers made on or after 15 March 1988 is charged at a single rate of 40%. For transfers made before 15 March 1988 there were sliding scales of tax as detailed in the Inland Revenue booklet IHT1 *Inheritance tax* obtainable from any of the Capital Taxes Offices mentioned at paragraph 3.21. Lifetime transfers which are chargeable when made are charged at half the rate of tax which would apply on death. Where such a transfer is made within 7 years of death, tax will become due at the full rate but will be tapered if that transfer occurred more than 3 years before the death. There are a number of reliefs and exemptions which are detailed in the booklet IHT1. In particular, most transfers between spouses are exempt.

Capital transfer tax

3.5 Capital transfer tax was charged on a sliding scale depending on the total value of chargeable lifetime transfers cumulated during the previous 10 years. There were separate scales for tranfers made on or within 3 years of death and for all other lifetime transfers. The reliefs and exemptions from capital transfer tax were broadly similar to those now ruling for inheritance tax.

The rules for woodlands

3.6 Woodlands are a chargeable asset but in addition to general exemptions detailed in the booklet IHT1 there are other reliefs which can be claimed in appropriate circumstances. These reliefs are explained in the following paragraphs.

Woodlands relief — deferment of tax

3.7 When a woodland is transferred on death an election may be made within 2 years of the death to have the value of standing trees and underwood, **but not the land on which they are growing,** left out of account in determining the value of the estate for inheritance tax (or capital transfer tax) purposes. When such an election is made the tax on the trees and underwood is deferred until they are disposed of by sale or otherwise. If however the beneficiary dies before all the timber is disposed of, the remaining deferred charge on the earlier death is wiped out, and a fresh election may be made to cover the timber which forms part of the beneficiary's estate at his or her death.

Conditions for relief

3.8 The woodlands must be in the UK and must not constitute agricultural property. For example small areas of woodland, the occupation of which is ancillary to that of agricultural land or pasture, do not qualify for the special relief for woodlands, although they may qualify for the separate relief given for agricultural property. If the woodlands were acquired by the deceased other than by gift or inheritance, he or she must have been beneficially entitled to them throughout the 5 years before his/her death. When woodlands transferred on death can be divided into **clearly distinct** geographical areas, elections may be made for each area separately. If woodlands are held jointly a joint election is necessary. The Inland Revenue regard an election, once made, as final.

Tax payable on disposal of timber

3.9 When the value of trees and underwood has been left out of account on a death and they are subsequently disposed of (other than to the spouse) tax becomes chargeable on the net proceeds of the sale if that disposal is for full consideration. In other cases, the tax is charged on the net open market value of the trees or underwood at the time of disposal. The rate of tax is found by adding the amount chargeable to the value of the estate of the last person on whose death the trees or underwood were left out of account and treating it as the highest part of that total. The scale in force at the date of disposal is then applied to the total. If there is more than one disposal the amount chargeable on a later disposal is added to the deceased's estate as increased by the amounts chargeable on earlier disposal(s). The disposal which triggers the deferred charge may itself be an occasion on which inheritance tax (or capital transfer tax where still applicable) is payable, for example if the timber is given away by the beneficiary. In that case, the value transferred is calculated as if the value of the trees or underwood had been reduced by the deferred tax chargeable on them by reference to the preceding death.

Admissible deductions

3.10 In calculating the net proceeds of sale or the net value of trees and underwood allowance is made for:

- the expenses of the disposal;
- the costs incurred in replanting, within 3 years of the disposal, the trees or underwood disposed of so long as they have not been previously allowed.

The Board of Inland Revenue has discretion to extend the 3-year period if the delay in replanting has been caused by circumstances outside the owner's control.

Business relief

General

3.11 The value of certain business property included in chargeable transfers during lifetime or on death may be reduced if the conditions outlined below are satisfied. The property to which the relief applies is described as 'relevant business property'.

Principal conditions for business relief

3.12 The business must be run on a profit making basis and the 'relevant business property' must have been owned by the transferor for a minimum of 2 full years immediately before the transfer. (For property inherited on the death of a spouse the period of the deceased's spouse ownership counts towards the 2-year period.) Or if the property (apart from a minority shareholding) replaced other 'relevant business property' the reduction is available if the two together were owned for a total of at least 2 out of the 5 years immediately before the transfer. Where a gift of a business becomes chargeable because the transferor dies within 7 years the recipient must still own that business — or qualifying replacement property — at the time of the transferor's death.

3.13 In the case of a woodlands business, business relief extends to the value attributable to the timber itself as well as to the residual value of the business or the interest in it. The relief also applies to the deferred charge triggered by a later disposal. It will apply to reduce the net proceeds or value of the timber in calculating the deferred tax charge if the conditions for business relief would have been satisfied at the time of the death on which the woodlands relief was granted. After 26 October 1977, business relief was also available in similar circumstances for capital transfer tax. Where an election is made for woodlands relief for the timber itself, business relief may be due on the value attributable to the rest of the business including the land on which the trees are growing.

Rate of reduction

3.14 The current rates of business relief are:

50% for
- a solely owned business or an interest in a business such as a share in a partnership;
- shares, whether quoted or unquoted, where the transferor had control of the company immediately before the transfer;
- unquoted shareholdings of over 25%;
- a life tenant's business or interest in a business;

30% for
- other unquoted minority shareholdings;
- land, buildings, machinery or plant owned by the transferor and used in his or her business;
- land, buildings, machinery or plant used by a life tenant in his or her business.

For full details you should refer to the booklet IHT1.

Heritage relief

3.15 Woodlands may qualify for the conditional exemption from inheritance tax (or capital transfer tax where still applicable) which is available for assets of national heritage quality. This exemption applies to land of outstanding scenic, historic or scientific interest as well as to certain other property. Trees and underwood may share in the exemption if they contribute to the qualifying interest. Ancient semi-natural woodlands which are, or could be, properly included on the Nature Conservancy Council's Inventory of Ancient Woodland will be eligible for consideration for exemption on scientific as well as scenic or historic grounds but each case will need to be considered on its merits. In return for exemption undertakings have to be given to preserve and maintain, and to provide public access to, the qualifying property. Tax becomes payable if the undertakings are broken or the property is sold or ownership changes (by lifetime gift or on death) and the relevant undertakings are not then given or renewed. Business relief is not available in these circumstances. The heritage relief provisions are described in detail in the booklet IR67 *Capital taxation and the national heritage* which is available from the Inland Revenue Reference Room, Room 8, New Wing, Somerset House, London WC2R 1LB (Tel: 01-438 6325), price £5.20. Cheques should be made payable to 'Inland Revenue'.

Timber used on the estate

3.16 Inheritance tax (or capital transfer tax where applicable) is not charged on timber cut for use on the estate. When an election is made to defer tax on underwood passing on death, a charge to tax is made only when the first subsequent cutting of underwood takes place.

Payment of tax

3.17 Inheritance tax (or capital transfer tax where appropriate) is due and payable within 6 months of the end of the month in which the transfer or event giving rise to the tax liability occurs, but in the case of lifetime gifts made between 6 April and 30 September inclusive, by the end of April in the following year. Personal representatives have to pay tax when they deliver their account of the deceased's estate (see paragraphs 3.21 and 3.22) but need not pay tax on woodlands if they elect for deferment (see paragraphs 3.7 and 3.8). When an election for woodlands relief has been made, the person liable to tax is the person who is entitled to the proceeds of the sale (or would be so entitled if the disposal were a sale). The deferred tax is payable within 6 months of the month in which the disposal occurs.

3.18 Except in the case of a lifetime gift where the donor is paying the tax (subject to the qualification below), or in the case of a deferred charge arising on a disposal of timber, the tax attributable to the value of a woodland business (including the land and timber comprised in it), or to the value of the land and timber on their own, may be paid by 10 annual instalments. The first instalment is payable on the date the tax falls due (see paragraph 3.17). If any of the timber is sold, however, the balance of the tax attributable to it becomes payable immediately. Generally, interest (currently at 11% but variable from time to time by Statutory Instrument) runs from the date when payment of the tax falls due. But instalments of tax on a woodland business are interest-free provided they are paid by the due date.

3.19 When a lifetime gift also triggers a deferred charge, the conditions relating to payment are relaxed for the tax payable on the gift. Tax may then be paid by instalments and the facility continues regardless of any sale. Also the instalments, including those for land and timber on their own, are interest-free if paid by the due date.

Transitional arrangements on the abolition of estate duty

3.20 Estate duty continues to be payable on the net proceeds of the sale of timber which was left out of account on a death before 13 March 1975 until the first transfer of value of the woodland (unless it is a transfer to a spouse). There are also special transitional arrangements for gifts made before 27 March 1974 where the transferor dies within 7 years (4 years in Northern Ireland) of the date of the gift.

Administrative arrangements

3.21 Inheritance tax is administered by the Capital Taxes Offices in London, Edinburgh and Belfast. Accounts and correspondence relating to estates taxable on death and to lifetime transfers should be sent to the appropriate Capital Taxes Office:

England and Wales
Minford House, Rockley Road, London, W14 0DF

Scotland
16 Picardy Place, Edinburgh, EH1 3NB

Northern Ireland
Law Courts Building, Chichester Street, Belfast, BT1 3NU.

Accounting arrangements

3.22 Where inheritance tax is payable on a death whether or not an election for deferment is made, an account should be delivered by the personal representatives within 12 months of the end of the month in which the death occured, or if later, within 3 months of the date on which the personal representatives begin to act. On the delivery of the account the personal representatives will have to pay any tax due. This will include tax on trees and underwood if an election for deferment is not made.

Part 4
Estate duty on woodlands

General

4.1 Estate duty was charged on the value of property, including settled property, passing on deaths between 1 April 1894 and 12 March 1975. For deaths up to 15 April 1969 the duty was charged at a prescribed rate based on the total value of the estate. After this date it was calculated on a sliding scale and an 'average' rate determined.

4.2 Where an estate included land on which timber, trees or wood was growing the value of the timber (but not the land) was left out of account in determining the value of the estate and the rate at which duty was payable. This relief did not extend to the timber value where the deceased had only a reversionary interest in the property. Where there is an outstanding deferred charge to estate duty, a transfer of value on or after 1 July 1986 involving the land bearing the timber cannot be treated as a PET (see paragraph 3.2). The result is that an immediate lifetime charge to inheritance tax will normally arise.

Estate duty payable on timber when sold

4.3 When any of the timber (other than underwood) which had been growing at the date of the deceased's death was subsequently felled and sold, the net proceeds — after expenses — were charged to estate duty at the rate which applied to the rest of the chargeable estate. The same applied where timber was sold standing and apart from the land, on the basis that the purchaser would fell it and pay for it as and when felled. In other cases where timber was sold standing estate duty became payable on its value as at the date of death, without any deduction for outgoings.

4.4 No duty was charged on the sale of casual windfalls or on timber cut for use on the estate. Also, the total value of the timber as at the death was in practice taken as the maximum figure on which estate duty would be charged by reference to that death, regardless of the total amount(s) actually realised on subsequent sale(s).

Accountability

4.5 The owners or trustees of the land were accountable for the duty as and when sale proceeds were received with interest (at the same rate as for outstanding inheritance tax — currently 11%) from the date of receipt.

Sale after 12 March 1975

4.6 The rules outlined above continue to apply to timber which last passed on a death before 13 March 1975, whether sold felled or standing, until the first transfer of value for inheritance tax occurs (see Part 3). Forms of account and any other information may be obtained from the appropriate address at paragraph 3.21 on page 9. Where the death occurred on or after that date, inheritance tax or capital tranfer tax (see Part 3) applies.

Value added tax

This Bulletin does not deal with the liability of woodland owners for value added tax (VAT), which is administered by HM Customs and Excise, New King's Beam House, London, SE1 9PJ.

ACKNOWLEDGEMENTS

The Forestry Commission is grateful to the Inland Revenue for assistance with the preparation of this Bulletin. The photograph *(22884)* used for the front cover illustration is from the Forestry Commission collection.

FURTHER READING

Information on forestry, planting, management, etc., is available from the range of Forestry Commission publications. A list of current titles is available in the Forestry Commission's *Catalogue of publications* obtainable free of charge from the Publications Section, Forest Research Station, Alice Holt Lodge, Wrecclesham, Farnham, Surrey, GU10 4LH (Tel. 0420 22255 ext. 305).